ON THE MOVE...

THE ALASKA PIPELINE

By Robert H. Redding

Consultant:
John Ratterman
Manager, Public Affairs
Alyeska Pipeline Service Co.
Anchorage, Alaska

CHILDRENS PRESS, CHICAGO

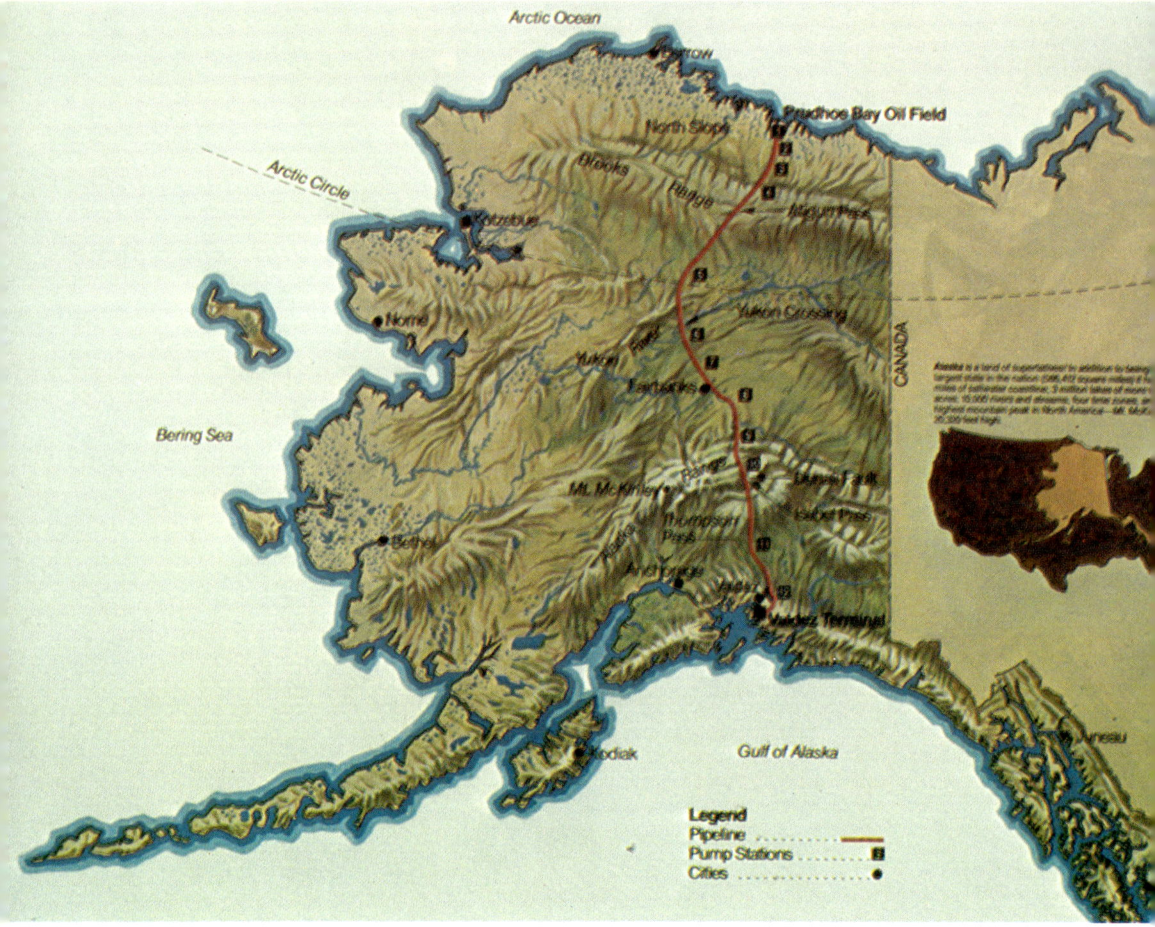

Library of Congress Cataloging in Publication Data

Redding, Robert H
The Alaska pipeline.

(On the move)
SUMMARY: Describes the construction and operation of the Alaska pipeline which runs from Prudhoe Bay on the Arctic Ocean to Valdez on Prince William Sound.
1. Alaska pipeline—Juvenile literature. [1. Alaska pipeline] I. Title.
TN879.5.R4 388.5 79-23150
ISBN 0-516-03883-4

1 2 3 4 5 6 7 8 9 10 11 12 R 87 86 85 84 83 82 81 80

Photographs courtesy of Alyeska Pipeline Service Co.

The Alaska pipeline runs from Prudhoe Bay on the Arctic Ocean to Valdez on Prince William Sound. It crosses frozen arctic tundra. It travels across the Brooks Range and the Alaska Range. It shoots across wide, wooded valleys and many rivers. The highest point, at 4800 feet, is Atigun Pass in the Brooks Range. From start to finish the pipeline is 800 miles long. That is a long way.

This is the Prudhoe Bay oil field. There are nine and four-tenths billion barrels of recoverable oil under the surface. That's a lot of oil. Some people think there is more oil than that. The Prudhoe Bay oil fields are located in Alaska's North Slope. This includes all the land north of the Brooks Range.

These are the pipe sections that were used in making the oil line. They are forty feet long, four feet in diameter, and weigh five tons each. They can withstand 1450 pounds per square foot of pressure.

You don't roll these pipes around like toothpicks!

The pipe for the pipeline was made in Japan. It was carried to Alaska by ship. One Japanese freighter, the *Alaska Maru*, was so large that it could haul eight miles of pipe in one trip.

Much of the pipe was welded to make lengths of eighty feet. It was hauled to staging areas on special trailers. In spite of the length of the pipe, there were few accidents. Good drivers saw to that.

The pipe was coated with a special material to keep it from rusting underground. Some of the pipe was green in color, and some gray. The color was a code to show how it was to be used.

Very modern methods of transportation were used to build the Alaska pipeline. It is one reason the line was finished in three years.

Some supplies and equipment were hauled by huge cargo planes. This is a Hercules C-130.

Some materials were hauled to job sites by sky crane helicopters. Others were brought by air cushion transporters.

BARGE 156
SEATTLE
BARGE 156
SEATTLE

Much of the pipe was hauled to Prudhoe Bay by barge. The trip was made once a year in mid-summer. But the Arctic Ocean's great ice pack was always a threat. Once a barge rolled over in the Arctic Ocean, and some of the pipe was damaged.

Sometimes barges carried entire buildings. Instant housing was needed. There were no lumber mills at the job sites.

A 360-mile road was built from the Yukon River to Prudhoe Bay, following the oil line route. The road took less than a year to build. It included 120 miles of access roads.

But it was a long, rough road. Workers had a saying that on this road the iron trucks were driven by rubber men and women.

There were nineteen camps along the route to house the workers. They bore strange names like Dietrich, Five Mile, Atigun, and Isabel. Each camp was a little city. Each had all the necessities: mess hall, bunk houses, recreation rooms, movies, and aid stations. From 150 to 3500 people lived in these camps at the height of construction.

They were the most modern workers' camps ever built.

Women worked at all kinds of jobs. There were 20,000 workers employed during construction. About 1900 of these were women. Most worked in offices, but many lived in camps. They worked as teamsters, laborers, surveyors, and iron workers. Jobs usually given to men.

Some even tried welding.

The workers on the pipeline often worked under harsh weather conditions. In Alaska temperatures range from 75 below in winter to 100 above in the summer. But no matter what the weather, the work had to be done.

Some of the pipe goes above ground, and some under water. Some was buried. About seven miles of pipe is refrigerated so it will not melt the permafrost and sink. Permafrost is permanently frozen ground. Alaska has plenty of it.

The entire pipeline when full of oil holds 9.04 million barrels of oil, figuring forty-two gallons per barrel. The oil travels at 130 degrees Fahrenheit, at seven and one half miles per hour. It takes four and one half days for the oil to travel the 800 miles.

Twelve pump stations push the oil through the pipeline. The pumps are of the gas turbine type, generating 13,500 horsepower each. There are three such pumps at each station. One is always on standby. It takes a lot of force to push a million gallons of crude oil on its way.

The pipeline is checked regularly. Airplanes check the line often. In addition, up-to-date electronic devices are used to spot leaks. They can spot leaks almost as soon as they happen.

Care has been taken so that the land and animals will not be harmed by the changes the pipeline made. The torn earth is being re-seeded. Arches have been built along the ancient migratory routes of caribou so the animals can pass under the pipeline.

No firearms were allowed construction workers. Some workers even made pets of the animals.

In Valdez, watch over the entire 800 miles of line takes place in a computerized control center. There are two controls, one in use, and one on backup status. The controls are operated twenty-four hours a day. If necessary, the computer controls can stop the entire oil flow in ten minutes.

This is where the 800-mile pipeline ends. There are 18 tanks at the Terminal. Each tank holds 510,000 barrels of oil. Each is 62 feet high, and 250 feet in diameter. The tanks are surrounded by special retaining walls of concrete and iron. In case a tank leaks, the wall will keep the oil contained. There are 52 miles of iron in a complete retaining wall.

The Terminal has three 430,000 barrel ballast tanks (the three nearest shore). Incoming ships must pump their ballast water into these tanks. The water is purified. When it is pumped back into the sea, it has only eight parts of oil to every million parts of water left. That isn't much.

Not so long ago, oil tankers pumped their ballast water directly into the sea. The sea suffered.

The docks have been built on bedrock. Valdez is earthquake prone, but the docks have been built to withstand big shocks. Oil tankers can discharge their ballast and be loaded with oil within three days. That is pretty fast!

The Alaska pipeline cost $8,000,000,000 to build, and took three years of actual construction. It took 20,000 people to do the job, but they did it! They built a modern engineering marvel, one that is capable of supplying the United States with twelve percent of its oil requirements.

INDEX

About the Author:

Robert Redding, a life-long Alaskan, lived in Valdez while the pipeline was being constructed. He was daily witness to the construction of Terminal Camp, and to line being laid in the area. Terminal Camp, across from Valdez, consists of the tank farm where the oil is stored while waiting shipment.

Mr. Redding is a published writer, with three books having been brought out under the Doubleday imprint. (*Aluk, the Story of an Alaskan Caribou*; *Mara, the Story of an Alaskan Weasel*; *North to the Wilderness*, autobiographical account of a youngster growing up in Alaska). He has been published serially in the respected *Alaska Magazine*, and has enjoyed acceptance by many periodicals both adult and juvenile, with emphasis on the juvenile. His stories for children have repeatedly reached wide circulation magazines such as *Ranger Rick's Nature Magazine*, *Wee Wisdom*, *The Friend*, and others for older age groups. He is, at present, working on another Alaskan book about experiences in wilderness living.